ANIMAL BATTL[...]

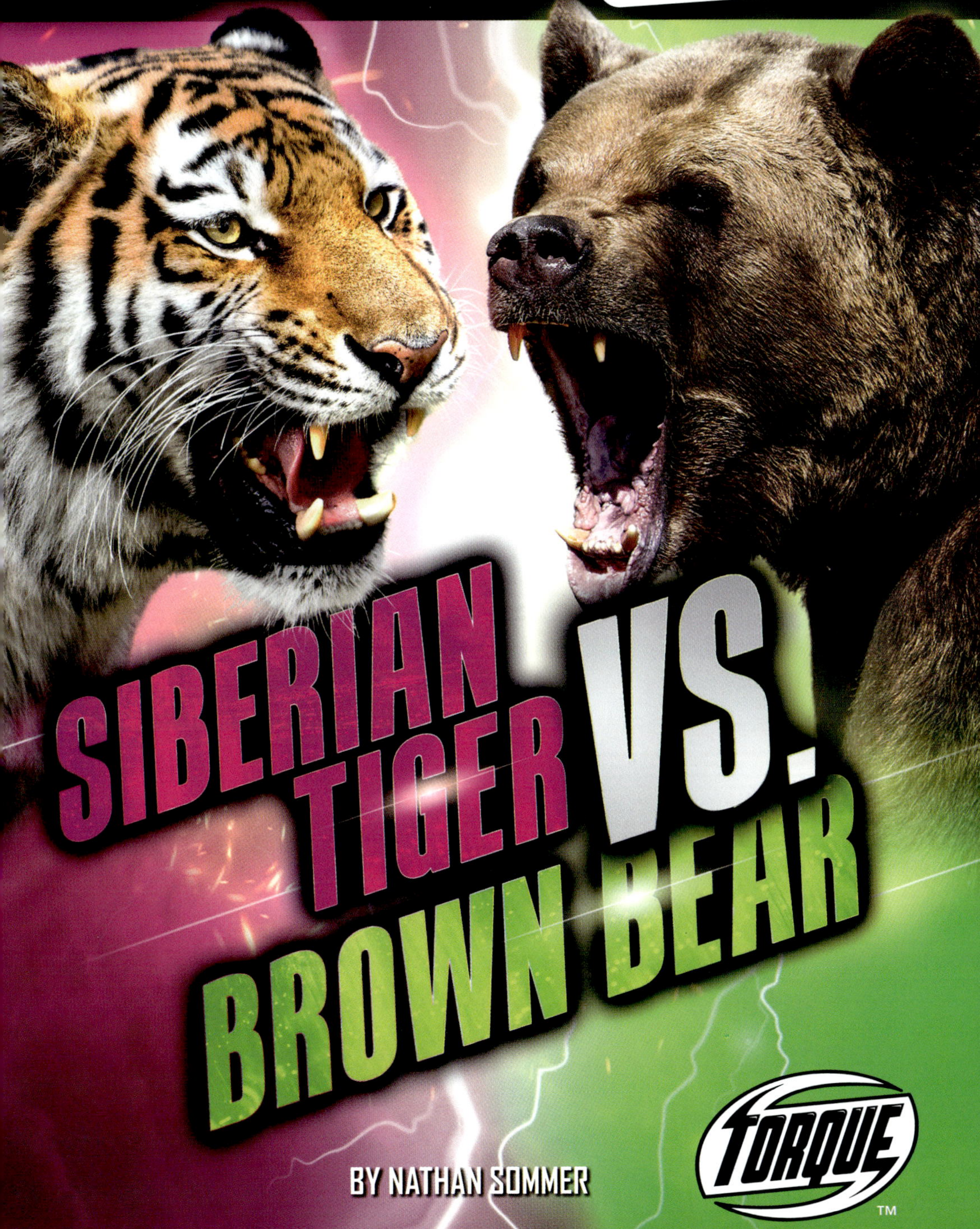

SIBERIAN TIGER VS. BROWN BEAR

BY NATHAN SOMMER

TORQUE™

BELLWETHER MEDIA • MINNEAPOLIS, MN

Torque brims with excitement perfect for thrill-seekers of all kinds. Discover daring survival skills, explore uncharted worlds, and marvel at mighty engines and extreme sports. In *Torque* books, anything can happen. Are you ready?

This edition first published in 2021 by Bellwether Media, Inc.

Library of Congress Cataloging-in-Publication Data

Names: Sommer, Nathan, author.
Title: Siberian tiger vs. brown bear / Nathan Sommer.
Other titles: Animal battles (Bellwether Media)
Description: Minneapolis, MN : Bellwether Media, 2021. | Series: Torque. Animal battles | Includes bibliographical references and index. | Audience: Ages 7-12. | Audience: Grades 4-6. | Summary: "Amazing photography accompanies engaging information about the fighting capabilities of Siberian tigers and brown bears. The combination of high-interest subject matter and light text is intended for students in grades 3 through 7"– Provided by publisher.
Identifiers: LCCN 2020003020 (print) | LCCN 2020003021 (ebook) | ISBN 9781644872826 (library binding) | ISBN 9781681037455 (ebook)
Subjects: LCSH: Siberian tiger–Juvenile literature. | Brown bear–Juvenile literature.
Classification: LCC QL737.C23 S5834 2021 (print) | LCC QL737.C23 (ebook) | DDC 599.756–dc23
LC record available at https://lccn.loc.gov/2020003020
LC ebook record available at https://lccn.loc.gov/2020003021

Editor: Kieran Downs   Designer: Andrea Schneider

Printed in the United States of America, North Mankato, MN.

TABLE OF CONTENTS

THE COMPETITORS

Eastern Asia's snowy landscape is one of the toughest in the world. The animals that live there must be strong to survive. The Siberian tiger is the king of this **habitat**

The big cats often compete with brown bears. These beastly bears are **apex predators** in most habitats. Who is the true champion of the forest?

RARE PREDATORS

Only about 500 Siberian tigers remain in the wild. This is due to habitat loss and hunting. Their body parts are viewed as trophies in some cultures.

Siberian tigers are the world's largest cats. They grow up to 13 feet (4 meters) long and weigh as much as 660 pounds (299 kilograms)!

These tigers live in eastern Russia's thick forests. They are also found in parts of China. The **solitary** creatures mark territories that they protect as their own.

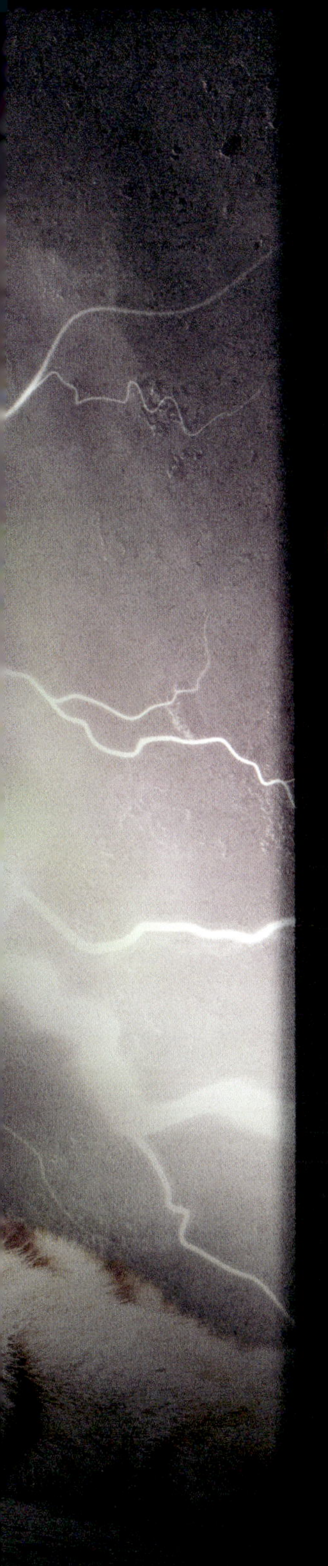

SIBERIAN TIGER PROFILE

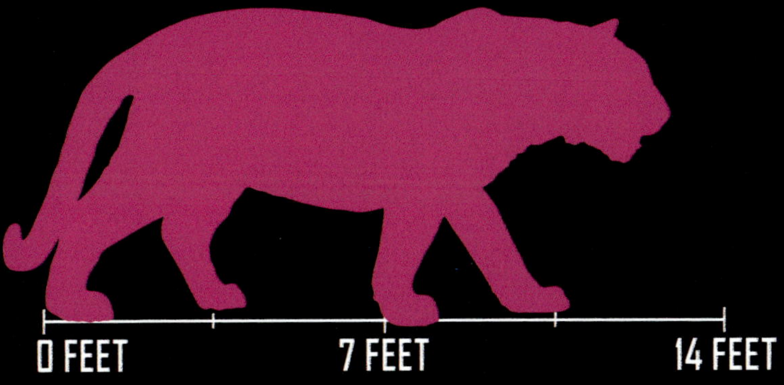

LENGTH

UP TO 13 FEET
(4 METERS)

WEIGHT

UP TO 660 POUNDS
(299 KILOGRAMS)

HABITAT

FORESTS

MOUNTAINS

SIBERIAN TIGER RANGE

BROWN BEAR PROFILE

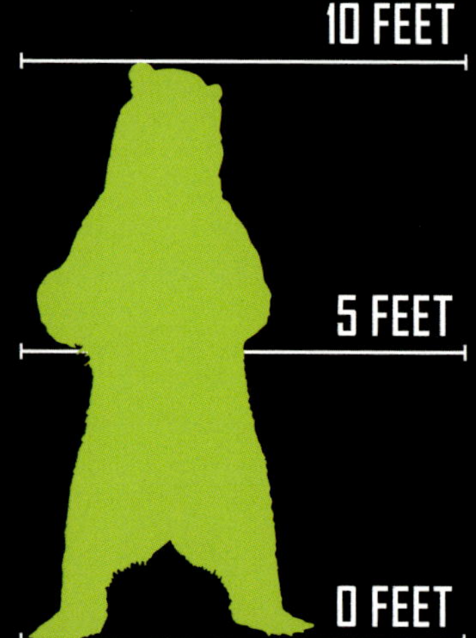

HEIGHT

UP TO 10 FEET
(3 METERS)

HABITAT

FORESTS

MOUNTAINS

WEIGHT

UP TO 1,500 POUNDS (680 KILOGRAMS)

BROWN BEAR RANGE

RANGE

Brown bears have furry bodies and large heads. They can be up to 10 feet (3 meters) tall when standing on two legs. These giants can weigh up to 1,500 pounds (680 kilograms)!

More than 80 **subspecies** of brown bears roam woodlands and mountains worldwide.

A LONG NAP

Brown bears sleep for 5 to 8 months over the winter. They do not even get up to eat or go to the bathroom!

SECRET WEAPONS

Brown bears are surprisingly speedy. They charge at speeds of up to 35 miles (56 kilometers) per hour when chasing **prey**. Not much escapes them!

Siberian tigers have special, thick bones in their front legs. These give them strength to take down large enemies. Shorter back legs let Siberian tigers leap more than 33 feet (10 meters)!

TIGER LEAPING DISTANCE

SIBERIAN TIGER
33 FEET (10 METERS)

LONG JUMP WORLD RECORD
29.36 INCHES (8.95 METERS)

0 FEET | 10 FEET | 20 FEET | 30 FEET | 40 FEET

SECRET WEAPONS

SIBERIAN TIGER

SHARP CLAWS

POWERFUL LEGS

SHARP TEETH

POWERFUL JAWS

Siberian tigers have sharp claws. Their long, curved shape makes grabbing prey easy. They can cut through skin in one swipe!

Brown bears have five sharp claws on each paw. The claws measure up to 4 inches (10 centimeters) long. They are used for both attacking and **defending**.

BROWN BEAR CLAW

4 INCHES
(10 CENTIMETERS) LONG

Siberian tigers have teeth with special **nerves**. They find the right part of an enemy's neck to bite. Powerful jaws then clamp down to **paralyze** prey!

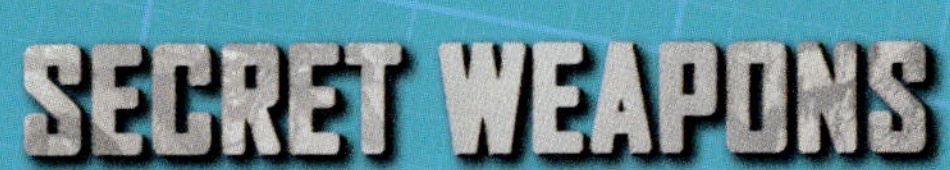

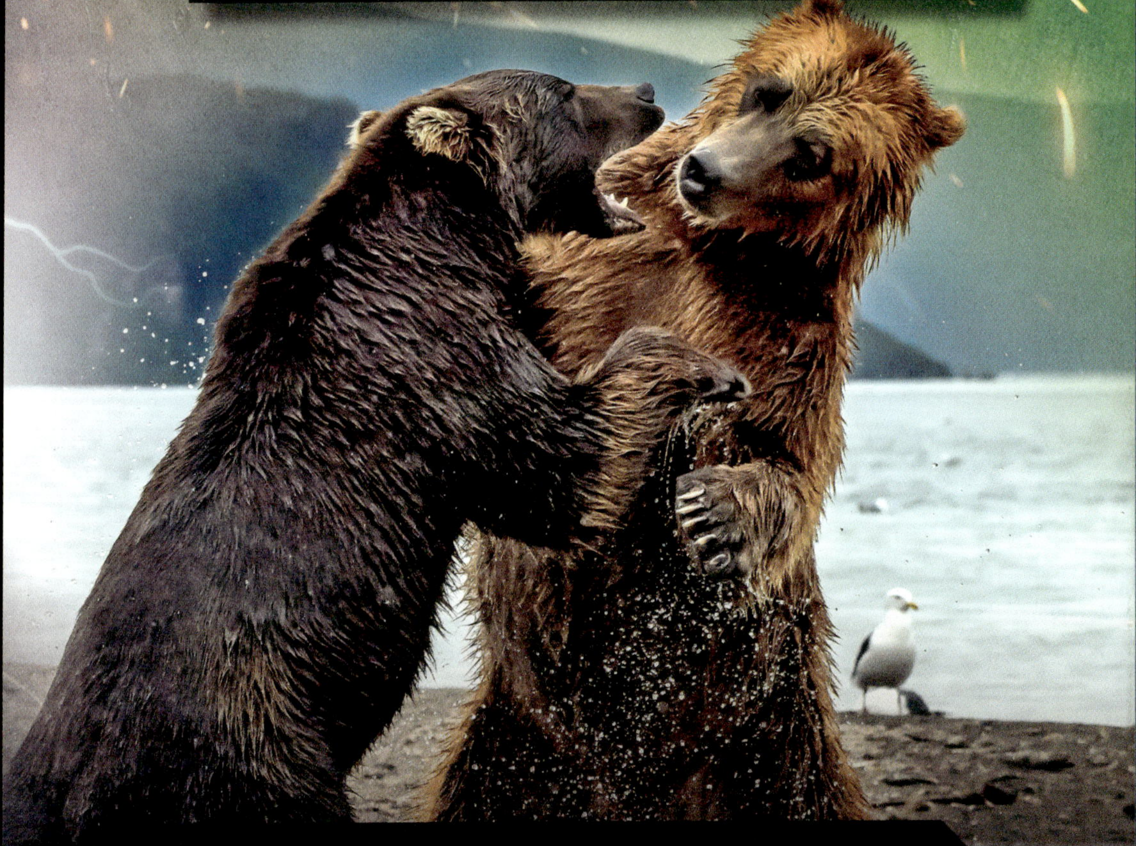

SECRET WEAPONS

BROWN BEAR

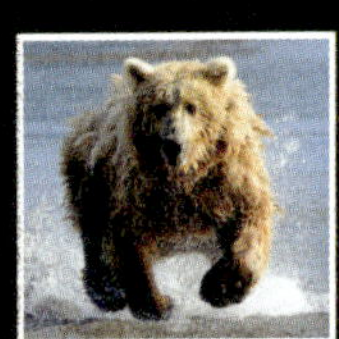

SPEED

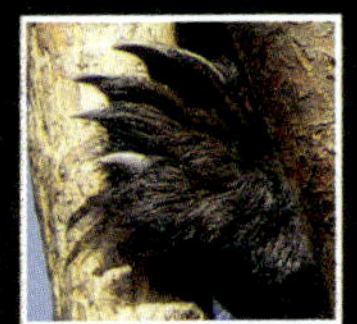

LONG CLAWS

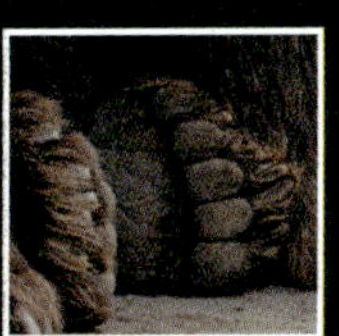

HUGE PAWS

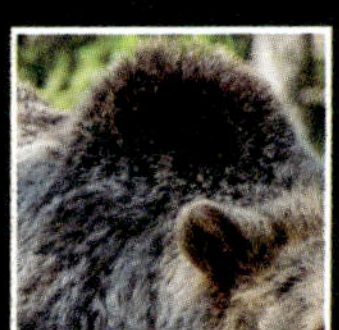

STRONG SHOULDERS

Humped, muscular shoulders make brown bears very strong. These shoulders, paired with huge paws, allow the bears to deliver big blows to attackers. One hit can bring enemies down!

ATTACK MOVES

Siberian tigers are very **aggressive** and **territorial**. They attack other animals that enter their territory. First, the tigers hide from enemies. Then, they attack from behind or the side!

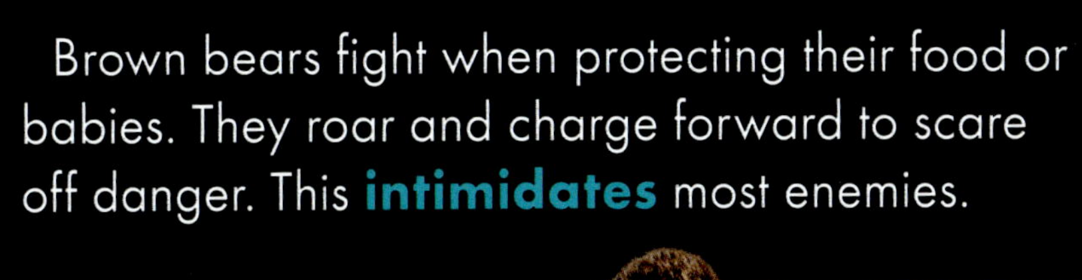

Brown bears fight when protecting their food or babies. They roar and charge forward to scare off danger. This **intimidates** most enemies.

SUPER SENSES

Brown bears can sniff out most enemies. Their sense of smell is hundreds of times better than a human's.

Siberian tigers pounce on enemies. They use their sharp claws to hold on. Then they attack with their sharp teeth. Most enemies cannot escape!

BLENDING IN

White patches of fur and black stripes help Siberian tigers blend into their habitats. Most prey do not notice them until it is too late!

Brown bears use paws and claws to bash foes if needed. This quickly takes down most enemies. Most animals do not stand a chance against these beasts!

READY, FIGHT!

A brown bear enters a Siberian tiger's territory. Sensing danger, the tiger hides until the bear walks by. The tiger pounces, but the bear quickly slams it down.

The bear swipes at the tiger. The tiger avoids the blow! It takes down the bear with a bite to the neck. Do not mess with this tiger's territory!

GLOSSARY

aggressive—ready to fight

apex predators—animals at the top of the food chain that are not preyed upon by other animals

defending—protecting

habitat—the home or area where an animal prefers to live

intimidates—frightens someone

nerves—parts of the body that send information to the brain from other parts of the body

paralyze—to make unable to move

prey—animals that are hunted by other animals for food

solitary—related to living alone

subspecies—particular types of animals that exist within a species

territorial—ready to defend a home area

TO LEARN MORE

AT THE LIBRARY

Daly, Ruth. *Bringing Back the Grizzly Bear.* New York, N.Y.: Crabtree Publishing, 2019.

Murray, Julie. *Tigers.* Minneapolis, Minn.: Abdo Publishing, 2020.

Sommer, Nathan. *Grizzly Bear vs. Wolf Pack.* Minneapolis, Minn.: Bellwether Media, 2020.

ON THE WEB

FACTSURFER

Factsurfer.com gives you a safe, fun way to find more information.

1. Go to www.factsurfer.com
2. Enter "Siberian tiger vs. brown bear" into the search box and click 🔍.
3. Select your book cover to see a list of related content.

INDEX

The images in this book are reproduced through the courtesy of: Thorsten Spoerlein, front cover (tiger); Michal Ninger, front cover (bear); GUDKOV ANDREY, pp. 4, 12; ArCaLu, p. 5; Shutterstock, pp. 6-7, 12 (sharp teeth); Sergey Uryadnikov, pp. 8-9, 15 (strong shoulders), 17; Petr Simon, pp. 10, 20-21, (tiger); Jan Stria, pp. 11, 12 (powerful legs); Michal Varga, p. 12 (sharp claws); HunsaBKK, p. 12 (powerful jaws); Design Pics Inc / Alamy Stock Photo, p. 13; Gerard Lacz, p. 14; Giuseppe_D'Amico, p. 15; AndreAnita, p. 15 (speed); Film Studio Aves, p. 15 (long claws); Azahara Perez, p. 15 (huge paws); Toshiji Fukuda, p. 16; slowmotiongli, p. 18; Kamilindoto / Alamy Stock Photo, p. 19; David Kalosson, p. 21 (bear).